NATIONAL GEOGRAPHIC

School Publishing

T0058621

Web Wizards

PIONEER EDITION

By Rebecca L. Johnson

CONTENTS

Keeping Watch. A female fishing spider guards a sac filled with many tiny spider eggs.

Web Wizards

Spiders are everywhere. They live in attics and basements. They live in forests and fields. Earth is home to about 40,000 kinds of spiders.

By Rebecca L. Johnson

What Are Spiders?

A spider is not an insect. A spider is an **arachnid**. A spider has eight legs. An insect has six. A spider's body has a head and an **abdomen**. That's its large back part. An insect's body has a head, an abdomen, and a third part, too.

Spider Parts. How is a spider's body different from a fly's body?

Super Silk

Spiders make thin strands called silk. Silk comes out of special structures on a spider's abdomen. These silk-making structures are **spinnerets**.

Silk shoots out of a spider's spinnerets. It is stretchy and strong as steel. Spiders can produce many kinds of silk for different jobs.

Spinning Silk. This Mediterranean black widow spider shoots silk out of its abdomen.

Tasty Treat. The wasp spider stings its prey before it eats it.

Ropes and Wraps

Spiders use silklike ropes. Some use silk to get down from high places. Jumping spiders attach a silk strand to a surface and leap. The silk strand saves them if they miss.

Female spiders make a silk egg sac. An egg sac may hold hundreds of eggs.

Wonderful Webs

Nearly all spiders are hunters. They catch and eat other insects and animals. Many spiders make webs. Webs are traps for insects and other small animals. Spider webs can be hard to see.

Webs have some smooth strands. Other strands are sticky. Small animals get stuck on the sticky strands.

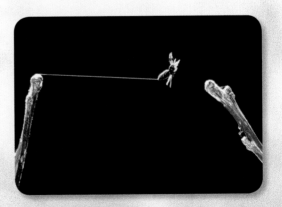

Tightrope Walker. This jumping spider uses silk strands to jump.

A Spider Life Cycle

A **life cycle** is all the stages in a living thing's life. A spider's life cycle starts with an egg. The egg is inside an egg sac. Some spider mothers stay with their egg sacs. Others hide them and leave.

Spider eggs hatch in a few weeks. A baby spider is a spiderling. A spiderling looks like its parents. Its features came from its mother and father. These features are called **inherited traits**. Eye color is an inherited trait. Something a spider does can be an inherited trait, too.

1

5

After their last molt, the spiders are adults. Each spider will lay eggs. Sh make an egg sac. The will hatch. The spider li is complete.

2 Spiders have a hard body covering. It doesn't stretch. Spiders must **molt** to grow. This means they replace their body covering with a new, bigger one. A new body covering forms under the old one. Then the old one comes off. Spiderlings molt in their egg sac. Then they come out.

3 Some kinds of spiderlings simply crawl away. Other kinds make strands of silk. The wind pulls the strands. The wind carries the spiderlings to new homes.

4 The spiderlings molt several more times. Each time they are bigger. They spin webs like their parents. Web-making is an inherited trait.

Wordwise

abdomen: the rear part of a spider's body

arachnid: an invertebrate animal with a hard body covering and eight legs

inherited trait: a feature or behavior passed down from parents

life cycle: all the different stages in the life of a living thing

molt: to get rid of an old body covering and replace it with a new, slightly larger one

spinneret: spider body part that makes silk

A Spinning Sampler

ifferent spiders make different webs. They move silk strands around with their legs. They work quickly. They are careful not to touch the sticky strands!

Sheet Web

Sheet webs are flat sheets of silk strands. The spider hangs below the sheet. Insects land on top of it. The spider pulls the insects down through the sheet.

Orb Web

An orb web looks like a wheel. Some strands come out from the center. It has sticky strands that go around and around.

Tangle Web

Tangle webs look messy. The strands go every which way. Look for tangle webs in high corners.

Funnel Web

Funnel webs are wide on one end. The other end is narrow. The spider hides in the narrow part. Insects get caught in the wide part. Then the spider grabs them.

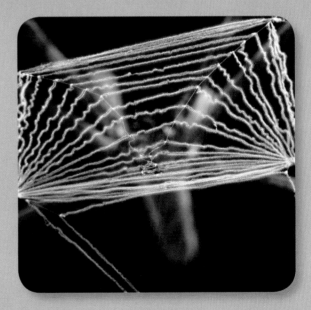

The Net

Net-casting spiders have long legs. They spin webs that they hold between their legs. Insects fly by. The spider tosses its net over them.

Veggie Spider!

Scientists thought that all spiders ate other animals. Then they discovered a small jumping spider in Central America. The spider eats plants.

The spider eats the tiny sweet tips of leaves. The leaves grow on a certain kind of plant. But ants guard it. They attack anything that tries to steal the leaf tips.

The spider watches the ants. Then it runs toward a leaf tip. The spider jumps away if ants move in. Or it leaps off the leaf and hangs on a silk strand. Usually, the spider gets away with a sweet leaf tip.

Standing Guard. This acacia ant is guarding the leaf tips on an acacia plant.

Web Wizards

Think you're a wizard when it comes to spiders? Prove it by answering the questions below.

1 What are the differences between a spider and an insect?

2 What are spinnerets?

3 How can a spider use silk?

4 How do young spiders know what kind of web to build?

5 Why is the veggie spider so special?